CHAINSAW OPERATOR'S MANUAL

CHAINSAW OPERATOR'S MANUAL

CHAINSAW SAFETY, MAINTENANCE AND CROSS-CUTTING TECHNIQUES

LAND
LINKS

LEARNING AND SKILL DEVELOPMENT
FOREST, WOOD, PAPER & TIMBER PRODUCTS INDUSTRY

National Library of Australia Cataloguing-in-Publication entry

Chainsaw operator's manual: chainsaw safety, maintenance and cross-cutting techniques / ForestWorks.

9780643097414 (pbk.)

Chain saws – Maintenance and repair – Handbooks, manuals, etc.

ForestWorks.

621.93

Published by
Landlinks Press
150 Oxford Street (PO Box 1139)
Collingwood VIC 3066
Australia

Telephone: +61 3 9662 7666
Local call: 1300 788 000 (Australia only)
Fax: +61 3 9662 7555
Email: publishing.sales@csiro.au
Web site: www.landlinks.com

Landlinks Press is an imprint of **CSIRO** PUBLISHING

Front cover photo courtesy David McElvenny

Set in 11/14 Adobe Gill Sans
Cover and text design by James Kelly
Typeset by Desktop Concepts Pty Ltd, Melbourne
Printed in Australia by Ligare

CSIRO PUBLISHING publishes and distributes scientific, technical and health science books, magazines and journals from Australia to a worldwide audience and conducts these activities autonomously from the research activities of the Commonwealth Scientific and Industrial Research Organisation (CSIRO).

Acknowledgements

Original author:
Bernard R Kestel
Manager OHS&R – Forests NSW

Technical advisors for this edition:
Trevor Ah Sam
 Auswide Projects
Russell Ainley
 NSW Forest Products Association Ltd
Ross Connolly
 TAFE NSW – Training and Education Support, Industry Skills Unit – Orange
Andy Cusack
 Logging Investigation & Training Association
Christine Di Bella
 Forests NSW
Greg Howard
 Timber Training Tasmania
Barry Levarde
 Barry Levarde Training, NT
Karl Liffman
 Timber Training Creswick
Barry MacGregor
 NSW Forest Products Association Ltd
Ian McLeod
 Director, McLeod Training Organisation Pty Ltd
 Safety Management Systems International Pty Ltd
David Priem
 Forest Industry Council
Bill Towie
 Forest Products Commission WA

Thank you to Husqvarna, Oregon Chain and Stihl for permission to use their material.

PEFC/21-31-17

The book has been printed on paper certified by the Programme for the Endorsement of Forest Chain of Custody (PEFC). PEFC is committed to sustainable forest management through third party forest certification of responsibly managed forests.

Contents

Introduction

A chainsaw is a portable power tool specifically designed for cutting wood. When used and maintained correctly it is very efficient. However, in the hands of inexperienced or careless operators, serious injuries and fatalities can occur.

Injuries to chainsaw operators have commonly resulted from a lack of saw control or from failing to follow safe work practices. Operator fatalities have occurred as a result of being struck by rolling logs, falling trees or dislodged tree limbs.

While improved technology and safety features on chainsaws, and the correct fitting of personal protective equipment, can help to reduce injuries, these offer little protection if the operator is not properly trained or does not follow safe operating techniques.

The major market for chainsaws now includes farming, local government, emergency services and recreational users who may not receive formal training. While this manual is designed primarily for all timber workers who use chainsaws in the course of their work, it will also provide guidance on safe chainsaw operation to the wider community.

The *Chainsaw Operator's Manual* was first published in 1983 by the Forestry Commission of NSW. Subsequent editions were: 2nd edition 1985, 3rd edition 1991, 4th edition 1993 and 5th edition 1997. The 6th edition was first published by Landlinks Press on behalf of Forests NSW in 2005. This new publication featuring chainsaw safety, maintenance and cross-cutting techniques but excluding tree felling was first published by Landlinks Press, © ForestWorks 2009.

This edition of the *Chainsaw Operator's Manual* includes updated information and additional material on chainsaw safety equipment and safe operating techniques.

A major change is the reorganisation of the manual into two separate books, reflecting the division of the content into basic chainsaw operations and manual tree felling operations.

Compliance with licensing, regulatory or certification requirements may be required in some states and jurisdictions. Please contact the relevant state authority listed at the back of this manual for current requirements.

1. National competency standards

The information in this manual can be used to support training aligned to the units of competency from the Forest and Forest Products Industry Training Package.

National competencies specify the skill and knowledge requirements for performing particular tasks or job functions in the workplace to the standard expected in the industry.

The units of competency listed below are partly covered by this manual. **Please note that completion of this manual does not constitute competence in these units.** Chainsaw operators seeking accreditation in any of the units below should consult a Registered Training Organisation (RTO). A list of RTOs can be found on the ForestWorks website given below.

FPICOT2004B	Maintain chainsaws
FPICOT2206B	Cross-cut material with a hand-held chainsaw
FPICOT2221B	Trim and cross-cut felled trees
FPIHAR2201B	Trim and cross-cut harvested trees

A copy of the units of competency can be downloaded from www.ntis.gov.au or www.training.gov.au.

For information and advice on learning and skills development please contact ForestWorks, the national skills advisory body for the forest, wood, paper and timber products industries: www.forestworks.com.au.

2. Know your chainsaw

Parts of a chainsaw

A chainsaw consists of two main units: the powerhead and the cutting attachment.

The powerhead

The powerhead is a high performance 2-stroke motor designed to work in any operational position at very high engine revolutions (typically in excess of 12 000 rpm). The motor is air cooled by a flywheel that forces the air over the cylinder cooling fins to dissipate heat.

The cutting attachment

The cutting attachment converts engine power into cutting performance. It consists of a guide bar, which attaches to the powerhead, and a loop of saw chain. The saw chain is driven around the guide bar by a drive sprocket. A centrifugal clutch provides the direct drive to the crankshaft.

The drive sprocket rotates at the same speed as the engine, resulting in chain speeds of up to 25 metres per second. As long as the chain is properly sharpened and correctly tensioned, it will self-feed into the timber with a minimum level of force required.

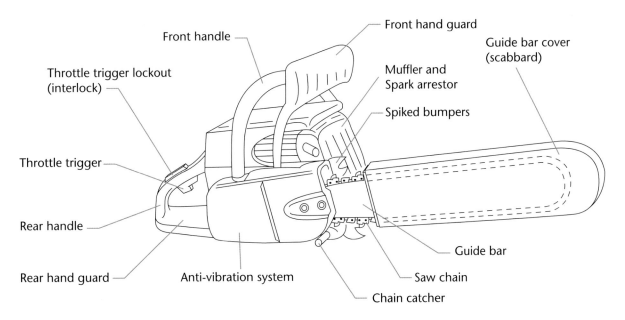

Figure 1: Components of a chainsaw.

Chainsaw safety features

Modern chainsaws are equipped with several features designed to improve operator safety. The operator must check that these devices are in good working condition before operating the saw. Fitted safety devices include:

- chain brake
- muffler and spark arrestor
- reduced kickback saw chain
- reduced radius guide bar
- chain catcher
- anti-vibration system
- hand guards
- decompression valve
- throttle trigger lockout (interlock)
- ignition switch
- guide bar cover (scabbard).

Chain brake

The chain brake may be internally mounted in the body of the saw, or externally within the sprocket cover. It is designed to stop a running saw chain in the event of a kickback reaction, and is used when starting the chainsaw.

Modern chainsaws are fitted with an inertia chain brake (ICB) which will automatically trip in any operational position should the saw react suddenly, or by contact with the front hand guard. This is especially important if the saw is being used with the guide bar in the horizontal position.

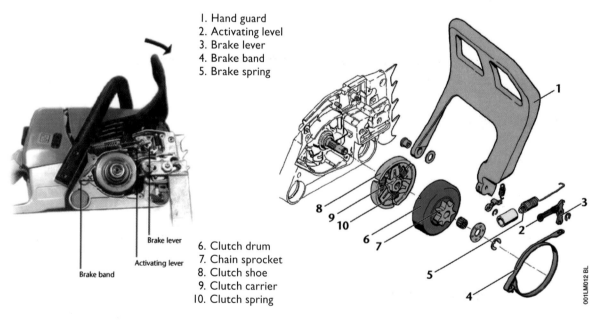

1. Hand guard
2. Activating level
3. Brake lever
4. Brake band
5. Brake spring

6. Clutch drum
7. Chain sprocket
8. Clutch shoe
9. Clutch carrier
10. Clutch spring

Figure 2: Chain brake mechanism.

Muffler and spark arrestor

The muffler is designed to reduce noise, and deflect exhaust gases away from the operator. The spark arrestor is a mesh screen that prevents spark emissions, which is especially important in the bush. Chainsaws with loose or damaged mufflers should not be used.

Reduced kickback chain

These types of chains are designed to ease saw chain cutter movement about the guide bar nose, thereby modifying kickback reaction.

Chain catcher

This is a device to intercept a broken or derailed chain and deflect it under the sprocket cover. It is located to the front of the sprocket cover, below the guide bar mounting. The chain catcher should be replaced if damaged.

Figure 3: Modern low kickback tendency chain. The ramp on the depth gauge guides the wood smoothly into the cutter.

Anti-vibration system

The anti-vibration system is designed to absorb the vibration created when the motor and chain are operating. It comprises a number of springs or rubber mounts that reduce the vibration reaching the operator through the handles. Excessive vibration can lead to nerve and circulation damage in the fingers, similar to Raynaud's disease or white finger disease.

Hand guards

Protection for the hands is provided by front and rear guards. The guard on the front handle protects the left hand and also serves to actuate the chain brake. The base of the rear handle protects the right hand when trimming and also in case the chain breaks or derails.

Ignition switch

The ignition switch is located where the thumb of the right hand can operate it while the operator is holding the rear handle. This design means the saw can be quickly shut off in an emergency.

Throttle trigger lockout (interlock)

Fitted to the rear handle, this prevents the accidental engagement of the throttle. The throttle cannot be depressed unless the mechanism is gripped by the operator.

Decompression valve

Activating this valve reduces the compression of the motor by venting the engine cylinder, thus lowering the effort required to start the chainsaw. Do not use the decompression button as a stop button as the valve could be damaged.

Guide bar cover (scabbard)

When not in use the chainsaw should be fitted with a guide bar cover (scabbard) to prevent injury to the operator or damage to the chain. The scabbard should be of sufficient length to cover the chain.

Reduced radius guide bar

Reduced radius guide bars are generally considered to be safer than those of broader profile since less circumference of the 'kickback reaction zone' is exposed during chainsaw operation. This zone is described as the quadrant of the guide bar from the very tip of the bar to the top radius of the bar nose.

Figure 4: Reduced radius guide bar (top).

Guide to selecting a chainsaw

There is a wide variety of chainsaw brands and models available on the market. Selection should be made after carefully analysing your needs and requirements. Consider the following points when evaluating the type of chainsaw required:

- What type of work will the machine be used for?
- What size material is expected to be cut on average?

These two questions will determine the saw capacity and guide bar length you require. The best indication of the power of the chainsaw is the capacity in cubic centimetres. You may also give consideration to the availability of service and spare parts, ease of maintenance, design features and balance of the chainsaw.

Table 1: Guide to chainsaw selection.

Classification	Power output	Engine capacity	Guide bar length	Main uses
Heavy	4.5 kW to 6.5 kW	80 cc to 125 cc	50 cm to 65 cm and longer	Timber falling. Sawlog preparation. Chainsaw milling.
Medium	3.0 kW to 4.5 kW	50 cc to 80 cc	To 50 cm	Small tree falling, trimming and cross-cutting.
Light	To 3.0 kW	To 50 cc	To 40 cm	Pruning.

3. Chainsaw safety

Operating a chainsaw can be a high risk activity. There are two Australian Standards that cover safety in chainsaw operations and that the operator should be familiar with:

AS 2726 Chainsaw Safety Requirements

AS 2727 Chainsaws – Guide for Safe Working Practices.

Develop a professional attitude

It is essential that a chainsaw operator develops a professional attitude towards all aspects of chainsaw use. Personal attributes that help to make a safe and competent operator include:

- forward planning and risk assessment
- putting safety first
- maintaining a steady work pace
- concentrating at all times
- using sound, low risk techniques
- understanding and taking care of your equipment
- caring about the environment.

Risk management

Using a chainsaw brings with it risks that the operator must be aware of, in order to take steps to minimise the possibility of accidents or injury. It is vital that you think before you act, and carry out a risk assessment of the work you are about to do.

Risk management is a process of identifying hazards in a workplace and eliminating or reducing as far as possible the risks associated with those hazards. There are four basic steps to risk management:

Table 2: Risk management strategy.

Step 1 Identify hazard	**Find out what hazards are present in the workplace.** *Example: Hanging limbs in trees overhanging the work area.*
Step 2 Assess risk	**Assess the risk – what could happen and why.** *Example: The hanging limbs could dislodge and fall on operators.*
Step 3 Control risk	**Evaluate and select options for minimising the risk.** *Example: 1. Prohibit work in the hazardous area* *2. Use mechanical equipment to remove the hanging limbs*
Step 4 Check risk controls	**Make sure the risk controls are working.** *Example: Ensure the area is cordoned off and operators kept out until hanging limbs are removed.*

Some of the things you need to consider before using a chainsaw include:

The operator

- Are you properly equipped and capable of doing the work?
- Do you have the necessary training and physical capabilities to do the work?
- Do you need assistance from other people, or special equipment?
- Are you tired, fatigued or under the influence of alcohol or medication?
- Do you have an adequate supply of drinking water?

The equipment

- Is your chainsaw sharp, properly maintained and suitable for the task at hand?
- Is it full of fuel and bar oil, so you don't need to stop the cutting sequence to refuel?
- Do you have other equipment such as an axe, wedges and hammer and first aid kit close at hand? Are they all in good condition?

The work environment

- What are the current and predicted weather conditions?
  Extreme temperatures, heavy rain, lightning or strong winds can make chainsaw operations very risky.
- What is the fire danger?
  Work should cease in the case of extreme fire danger. In summer months, keep fire fighting equipment handy.
- Are there any physical ground hazards that might hinder your movements?
  Look for rocks, undergrowth, stumps, holes, etc.
- Are there any unsafe trees nearby?
  Look for stags, hang-ups or burnt out trees. **Look up!** Are there any hanging dead or broken limbs overhead? (Known as 'widow makers' for obvious reasons.)
- Who else is in the area?
  Be aware of other public activity, traffic or machinery movements.
- What are your workmates doing?
  Is there sufficient separation distance between the saw operator and other personnel? Do not work alone, and maintain regular contact with other workers.
- Is there tree felling activity in the work area?
  Do not work within at least two tree lengths (minimum) of any tree felling activity.

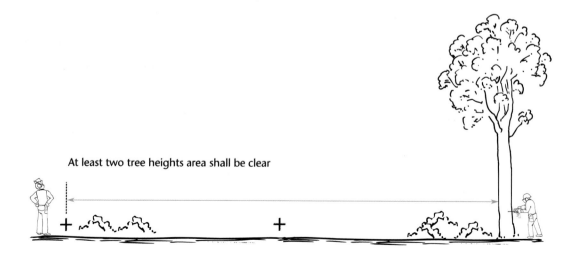

At least two tree heights area shall be clear

Figure 5: Area to be isolated during falling activity.

Personal protection equipment

The Australian Standard *AS 2727 Chainsaws – Guide for Safe Working Practices* lists the items of personal protection equipment (PPE) that a chainsaw operator must use. Remember that PPE is at the lower end of the scale of risk control, and will not guarantee your safety. Proper risk assessment and safe work practices must always be followed.

The items below are recommended when working in a forest environment. Refer to the list of technical standards at the back of this manual for more detailed specifications.

- Safety helmet (AS/NZS 1801)
  Must be replaced if cracked, damaged or past expiry date. Avoid damage caused by attaching stickers, storing in direct sunlight and contact with solvents. A legionnaire-style flap can be attached to protect the back of the neck.

- Eye protection (AS/NZS 1336 and 1337)
  Preferably non-scratch and non-fogging. Can be either clear or mesh visors or goggles. Safety glasses should also provide an acceptable level of protection.

- Steel-capped safety boots (AS/NZS 2210)
  Boots should also have non-slip tread and lace-up for better ankle support.

- Cut resistant trousers or chaps (AS/NZS 4453)
  The cut resistant layers cannot be repaired and the garment should be replaced if it has been cut. The effectiveness of the cut resistant layers may be reduced over time by the absorption of oil when used regularly.

- High-visibility vest/shirt (AS/NZS 4602)
  Long sleeves are preferable for sun protection. Some vests are also reinforced with cut resistant fabric for added protection.

- Safety gloves
  Should be snug fitting and of a hard wearing, protective fabric. Some gloves have gel cushioning to protect against vibration. Protective wrist bands and forearm guards are also available.

- Hearing protection (AS/NZS 1270)
  May be either ear plugs or ear muffs. Be aware that ear plugs and some cheap ear muffs may not provide sufficient protection when using larger saws. Check with your dealer for correct level of protection required for your chainsaw.

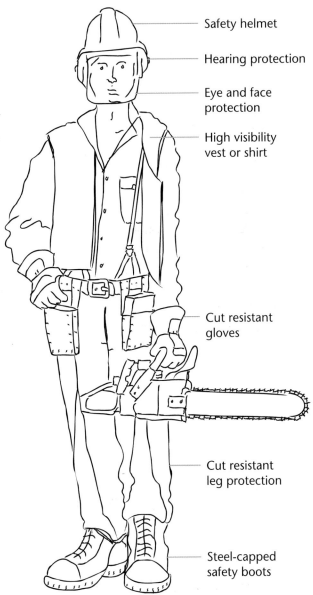

Safety helmet

Hearing protection

Eye and face protection

High visibility vest or shirt

Cut resistant gloves

Cut resistant leg protection

Steel-capped safety boots

Figure 6: Personal protective clothing and equipment.

Other equipment

Some tools and items of equipment are also considered necessary for timber workers when performing chainsaw operations.

Sharp axe

This can be used for many tasks, such as removing bark from a log. The axe should be in good condition, with a forged and tempered head free of cracks and a sharp cutting edge. When not in use a protective stitched leather cover should be fitted.

The best handles are made of American hickory or Australian hardwood timber. These should be tightly fitted, secured with cross-wedges and/or pinned. Ideally a handle should be straight grained, generally smooth and free of cracks or knots, and have a roughened grip. Linseed oil may be applied to keep the handle in good condition.

Wedges and hammer

At least two wedges are required. These should be made of robust aluminium alloy or lightweight plastic that will not adversely damage a running saw chain if struck. Wedges must be in good condition, free of cracks and mushroomed heads. While not recommended for general use, steel wedges may be useful for fire salvage work.

A suitable hammer of sufficient weight is required for driving wedges. The handle must be of good length, tightly fitted and free of knots, cracks and splinters.

Figure 7: Axe and cover.

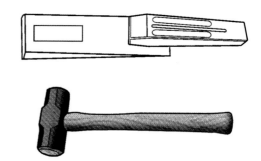

Figure 8: Wedges and hammer.

Cant hook or breaking bar

A cant hook or breaking bar can be of assistance in moving log material and bringing down hung up trees.

Fuel and oil container

Fuel and oil containers for storage and dispensing of 2-stroke mix and chain oil must be of approved fuel-proof design. Some have 'quick-fill' automatic cut-off fuel pourers.

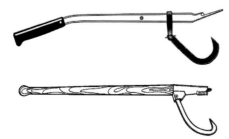

Figure 9: Cant hook and breaking bar.

Figure 10: Fuel and oil container.

First aid kit

A fully equipped first aid kit must always be provided at the workplace, and a personal kit must be readily accessible by the operator. Among the contents will be large dressings for lacerations. Sun protection cream must also be available as necessary. The Australian Standard AS 2727 recommends the minimum content of the kit.

Fire control equipment

Fire control equipment including a rake-hoe, fully charged hand pump knapsack-type sprayer and fire extinguisher are required during the fire season.

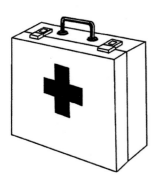

Figure 11: First aid kit.

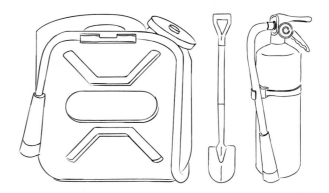

Figure 12: Fire control equipment.

Tool belt

A tool belt is worn for easy carrying of logger's tape, wedges and such small tools as a chainsaw wrench, file and possibly a log vice. A small first aid kit can also be attached.

Figure 13: Tool belt with logger's tape attached.

Logger's tape

Logger's tape is a specially designed retractable, spring loaded steel tape, necessary for measuring saw logs when cross-cutting to length.

Warning signs

Workplace signage needs to be erected at all entry points to the work area. Workplace signage warning of falling trees must be prominently erected beside coupe access roads and tracks.

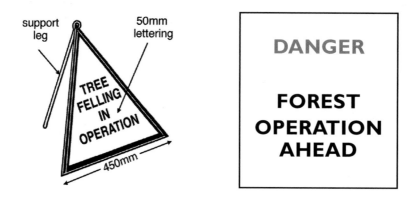

Figure 14: Workplace warning signage.

Chainsaw maintenance kit

The most basic chainsaw maintenance kit is that carried on the operator's tool belt. A comprehensive kit contains the following small tools:

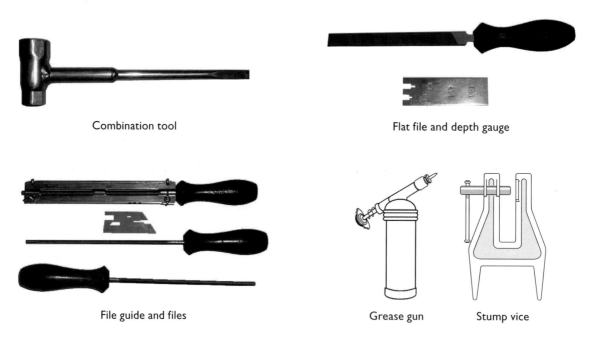

Figure 15: Some components of a basic maintenance kit.

Additional items include wrenches, a flat head screwdriver, wooden scraper, brushes, air filter treatments and cleaning rags. It is essential that all tools and equipment are well maintained in sound condition.

4. Chainsaw maintenance

Regular maintenance of the chainsaw, in accordance with the manufacturer's advice, is essential for safe, reliable and trouble-free operation. Chainsaw breakdowns are costly both in repair expenses and lost production. A professional attitude to maintenance means paying attention to:

- the recommended frequency of servicing
- careful inspection of all components
- cleanliness of the saw and related equipment
- reporting of problems and faults promptly
- arranging for specialist repairs when required.

Do not operate the saw if any components become defective, particularly any safety devices.

Powerhead and guide bar maintenance

Daily service (or frequent service as required)

The maintenance schedules are based on a regular daily engine operating time of 6 to 8 hours. However, when working in particularly dirty or dusty conditions, or when cutting dry timber, the service intervals may need to be shortened. The tasks below should be carried out daily according to manufacturer's specifications:

Cleaning
- Thoroughly clean saw body, air intake vents and cooling fins.
- Use plastic or wooden scrapers, not metal screwdrivers which can damage paintwork, leading to corrosion.

Air filter
- Before removing filter, apply the choke to close off the carburettor throat to prevent the entry of foreign matter. Remove and check the filter for damage and penetration of foreign material. Clean as per manufacturer's recommendations. Replace filter if damaged.
- A spare filter is useful to have as a replacement as required.

Figure 16: Checking the air filter.

Chain brake

- Thoroughly clean, particularly around brake band and operating mechanism.
- Frequent operation of the brake throughout the day will keep the brake's internal components free from an accumulation of dirt and saw dust.
- Ensure correct operation.

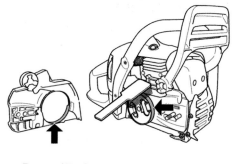

Figure 17: Cleaning the chain brake.

Saw chain and guide bar

- Remove side cover, clean and inspect.
- Remove saw chain and visually inspect chain and bar for wear and cracks.
- Clean out groove, working from the nose backwards. Remove any burrs.
- Clean oil holes.
- Sprocket nose:
  - clean out debris
  - ensure sprocket rotates and check for broken teeth
  - grease, if appropriate.

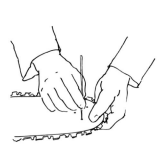

Figure 18: Cleaning and greasing the sprocket nose.

- Reverse the guide (cutter) bar to equalise wear.
- Install chain and sharpen, or fit spare chain. It may be necessary to change or sharpen the chain several times during the day.
- Tension chain correctly.

Loose screws/nuts

- Check screws/nuts for tightness. Do not over-tighten.

Safety features

- Check all safety devices are operational.

Fill fuel and chain oil tanks

- Filling the tanks at the end of the day will minimise condensation of water in the tanks during storage.

Periodic service

In addition to the daily service items, refer to the owner's manual for other tasks that should be carried out on a regular basis. This may include:

Drive sprocket
- Check weekly for wear.
- Replace after two chains or sooner if wear marks on teeth exceed 0.5 mm.
- Always clean crankshaft stub and needle bearing, and grease bearing when replacing sprockets.

Spark plug
- Remove, clean and check gap. Replace as required.

Fuel system
- Filter – check.
- Fuel tank – flush out any accumulated saw dust with 2-stroke mix.

Chain brake
- Frequent lubrication of the pivot or sliding surfaces is necessary to ensure effective operation.
- For chain brake adjustment, refer to a competent chainsaw mechanic.

Oil system
- Check operation, clean as necessary.

Cooling fans
- Remove fan housing, clean fan and cylinder fins.

Anti-vibration system
- Check springs/rubber mountings for looseness and deterioration.

Spark arrestor
- When engine performance begins to deteriorate, remove and clean spark arrester. The spark arrester must be securely fitted at all times.

Tune engine
- Most modern chainsaws require a digital tachometer for tuning and it is recommended to refer to a dealer/service person.

Starter cord
- Check the starter cord for wear, and replace to prevent breaking on the job. The most common point of cord wear is within the first 20 cm.

Table 3: Chainsaw powerhead troubleshooting chart.

Problem	Possible causes	Remedy
Motor starts, runs well but loses power after two or three cuts. It will pick up again after a brief rest but the condition worsens rapidly.	• Motor overheating due to: too lean a mixture; not enough oil in 2-stroke fuel • Pre-ignition by small metallic particles on spark plug	• Tune motor; obtain correct mixture • Replace spark plug
Motor starts and runs well but will not continue to idle with throttle in idle position. Speed of motor gradually increases until motor is racing.	• Leaking fuel pump diaphragm gasket allowing petrol to leak through pump impulse channel into crankcase • Worn throttle shutter and shaft	• Replace gasket. If fuel pump body is warped, replace. • Replace shutter and shaft
Motor runs well but loses power at certain angles.	• Carburettor inlet control lever is incorrectly set	• Reset lever so it is level with body of carburettor
Excessive vibration.	• Anti-vibration mounting broken • Depth gauges on chain too high • Chain not filed correctly • Worn sprocket • Chain pitch and sprocket pitch not matched • Loose clutch or loose fly wheel	• Replace • Re-set • File carefully • Replace • Use correct chain, sprocket • Refer to dealer
Motor will not start.	• No fuel in tank • Motor flooded (remedy a or b) Other possible causes: • Faulty spark plug • High tension lead shorting due to dampness or wear • Ignition switch off or faulty • Jets wrongly adjusted • No compression	• Refuel • a) Place saw over log with exhaust port facing down and choke in off position. Hold throttle wide open and pull starter. • b) Set choke to running position, turn off ignition. Remove and dry out spark plug. Ensure spark plug lead is earthed and briskly pull starter rope to blow out fuel. Replace spark plug.
Motor starts and misses.	• Faulty carburettor adjustment • Faulty spark plug • Leaking or damaged high tension lead, damp • Faulty on/off switch • Faulty electronic ignition	• Readjust • Replace • Replace or dry out • Replace • Replace

Table 3: (Continued)

Problem	Possible causes	Remedy
Motor starts, will not idle but runs at very high speed.	• Low speed jet wrongly adjusted	• Re-adjust
Motor starts, idles but will not accelerate to full speed.	• Chain adjusted too tightly • Fuel filter plugged with dirt • Chain drive links jammed by pinched bar groove • Choke shutter not opening • Low or high speed jet wrongly adjusted • Blocked exhaust screen	• Re-adjust • Clean or replace • Refer to dealer • Replace • Re-adjust • Clean or replace
Motor starts, idles and appears to run well, but lacks power.	• Tight chain • Jammed chain • Exhaust ports clogged with carbon • Dirty air filter element • Blocked fuel filter • Incorrect tuning • Rings and/or cylinder worn • Worn or inoperative clutch	• Re-adjust • Free • Bring piston to top dead centre. Clear carbon from port using wood scraper. • Clean or replace • Clean or replace • Check and adjust • To test: remove starter assembly and turn motor by hand in forward and reverse direction. Higher compression in reverse indicates worn rings and/or cylinder. • Replace
Motor idles but chain continues to move.	• Motor idling too fast and chain too loose • Clutch drum bearing worn, dry • Clutch springs broken • Clutch shoe pivot point worn • Clutch shoes broken	• Re-adjust idling, tension chain • Oil or replace • Replace • Replace • Replace
Motor idling speed can only be obtained with mixture settings very different from specification.	• Damage to adjustment screws by screwing too tight. Jet seats likely damaged also. • Worn throttle shaft and bushing	• Replace with new screws and new carburettor • Replace shaft and bushing

The cutting attachment

The cutting attachment on a chainsaw does the real work when the saw is operating. Therefore it is essential that all three components – drive sprocket, guide bar and saw chain – are properly maintained for the saw to be used efficiently and safely. The drive sprocket, guide bar and chain work as a closely related trio. If any of these parts are not in top working order, problems will result in the other two. For example, a loose chain will cause rapid wear behind the nose of the bar and will increase wear on the guide bar.

Drive sprockets

The clutch transmits power through the clutch drum to the drive sprocket. The sprocket draws the chain around the guide bar and through the wood. The drive sprocket is a gear, and gear terminology is used to name its parts. There are two types of drive sprockets: spur (fixed) and rim (floating).

Spur sprockets

Spur sprockets have wide tooth faces which engage the chain's drive link tangs. These wide faces help in the alignment of the chain to the guide bar groove. The spur sprocket supports the chain either on the tip of the teeth or by nesting the drive link tang between the sprocket teeth. Spur sprockets are permanently attached to the clutch drum. They have the advantages of good bark and chip removal but tend to cause more wear on the saw chain than rim sprockets. These sprockets are also more expensive to replace.

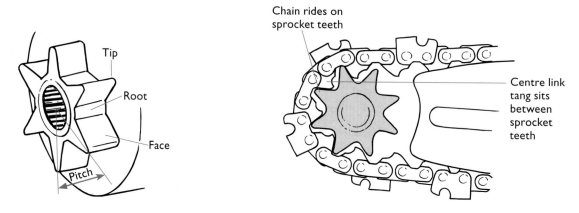

Figure 19: Spur (fixed) drive sprocket.

Rim sprockets

Rim type sprockets look like a wheel because they have the sprocket teeth mounted between two rims. The rims support the tie straps and cutters thus allowing the chain to run more smoothly with less wear. The drive link (tangs) of the chain is engaged by the faces of the sprocket teeth.

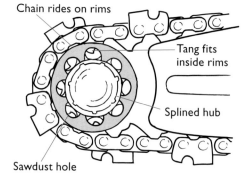

Figure 20: Rim (floating) drive sprocket.

Rim sprocket fitting

The rim sprocket is mounted on a splined hub, welded to the clutch drum. This spline allows the rim sprocket to self align with the guide bar groove.

The more common type of rim is now the radial port rim. This sprocket uses a system of radial ports to discharge the wood chips. The discharge holes face away from the clutch drum.

Drive sprocket replacement

Drive sprockets are subjected to extreme heat and friction and therefore wear. A worn drive sprocket will damage and weaken a chain beyond repair, cause a loss of cutting power and accelerate guide bar wear.

Some manufacturer's recommend installing a new drive sprocket with each new chain fitted. However, drive sprocket life can be extended by using two chains per sprocket with chains being alternated daily.

To avoid drive sprocket problems:

- replace when worn with a correctly pitched sprocket
- keep chain well sharpened and correctly tensioned
- grease bearing each fortnight or upon drive sprocket removal from saw
- keep guide bar rails in good order
- maintain a good oil flow in guide bar groove.

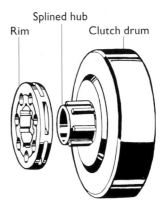

Figure 21: Fitting a rim drive sprocket.

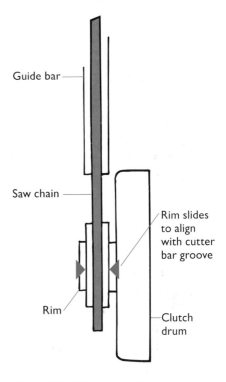

Figure 22: Alignment of drive sprocket.

Figure 23: Worn drive sprocket.

Drive sprocket pitch

It is important that the pitch of the chain, drive sprocket and bar nose all match, so the chain drive links fit neatly into the sprocket. The pitch of the chain is determined by measuring the distance between the centres of three adjacent rivets, and dividing by two.

The pitch size is described as an imperial or metric fraction of an inch.

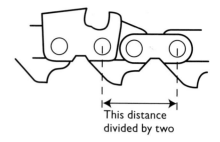

This distance
divided by two

Figure 24: Measuring the pitch.

Drive sprocket size

Chainsaws usually come fitted with 7-tooth sprockets which are designed to give the most useful combination of high chain speed and high torque in most cutting situations.

Fitting an 8-tooth sprocket will increase chain speed but decrease torque, which is an advantage for trimming small limbs or cutting wood significantly smaller in diameter than the length of the bar.

Guide bar

The guide bar provides a track for the chain to run on.

Guide bar types may be solid nose or sprocket nose. In general, sprocket nose bars are preferred as they provide safer, cooler and more efficient operation.

The **solid nose bar** has a protective layer of stellite around the nose of the bar to prevent wear.

Solid nose profiles are generally of wide radius, and are useful in dirty, high contamination situations.

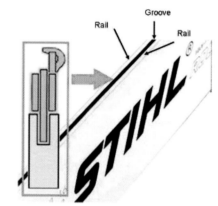

Figure 25: Guide bar.

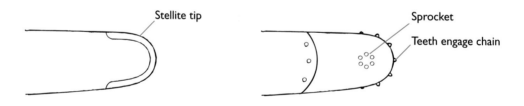

Figure 26: Solid nose and sprocket nose guide bars.

Sprocket nose types include laminated, replaceable and narrow profile.

A sprocket nose guide bar allows the saw chain to run more tightly than a solid nose bar. The sprocket reduces friction, thus allowing more power to be delivered to the saw chain.

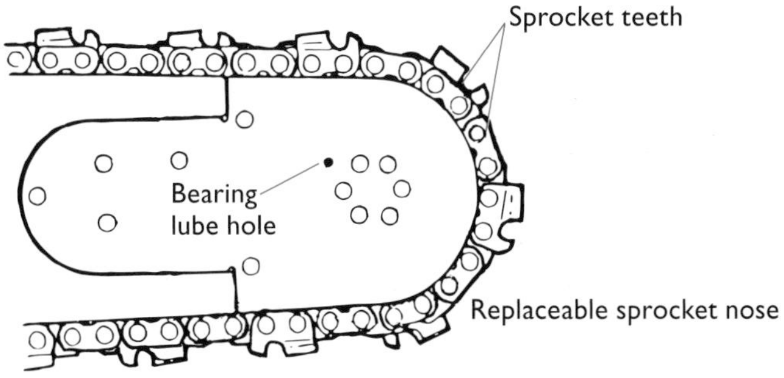

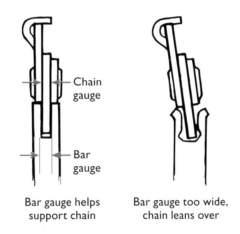

Figure 27: Replaceable sprocket nose guide bar.

Guide bar maintenance

- Keep the groove clean and maintain the clearance to the bottom of the chain drive links at minimum 1.0 mm (1/32").
- Remove feathered or burred edges with a flat file or stone. File from the outside in.
- Check the oil holes regularly to ensure they are clear.
- Turn guide bar daily to ensure even wear.
- Ensure chain entry point is properly funnelled (lead-in).
- Ensure the rails are even and run at right angles to the body of the guide bar.

Chain gauge

Bar gauge

Bar gauge helps support chain

Bar gauge too wide, chain leans over

Figure 28: Ensure correct guide bar gauge.

Figure 29: Cleaning the groove.

Figure 30: Removing burred edges.

Saw chain

Chains come in many different types, shapes and sizes, but all chains have common requirements for efficient operation:

- They need to be sharpened correctly and regularly.
- They need to be well lubricated.
- They need to be tensioned correctly.
- Their depth gauges need to be set correctly.

Chain components

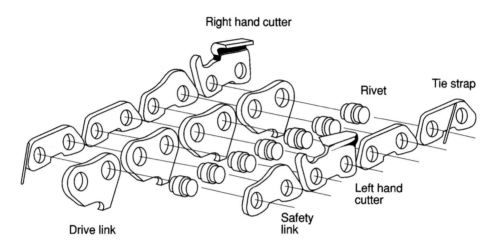

Figure 31: Exploded view of saw chain components.

Saw chain design

In the early 1980s, chain makers designed a ramp into the tie strap between the cutters to lead the wood into the cutter. This design was referred to as the 'first generation safety chain'. The original design was further progressed to include a ramp in the drive link and then the depth gauge.

Today's chain manufacturers market a range of saw chains designed to satisfy particular functions, such as low vibration, reduced kickback, abrasion resistance, and those for specialised applications (e.g. heavy bark and ripping).

Several different designs of chain profiles have come onto the market in recent years, so seek guidance from your chainsaw dealer regarding the most suitable chain for your requirements.

Figure 32: a) Standard non-safety chain. The wood strikes the vertical face of the depth gauge in an aggressive manner.

b) First generation safety chain with raised ramp on tie straps.

c) Modern saw chain. The ramp on the depth gauge guides the wood smoothly into the cutter.

Cutters

A saw chain has both left and right cutters.

Each cutter has a top plate and side plate which both need to be sharpened to a fine edge.

The top plate feeds the cutter into the wood and the side plate severs the side of the cutting track.

The ramp in front of the cutting edge is called a depth gauge. This leads the cutter into the wood and determines how large a 'bite' the cutter will take (measured by the difference in height between the top of the depth gauge and the height of the top plate).

If the depth gauge is too high, the chain will not cut efficiently. If the depth gauge is too low, the cutter will grab too much wood and become jammed, and may cause kickback.

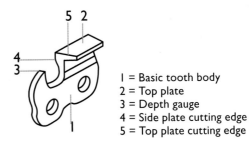

1 = Basic tooth body
2 = Top plate
3 = Depth gauge
4 = Side plate cutting edge
5 = Top plate cutting edge

Figure 33: Parts of a cutter.

Cutter profiles

Chipper profile (round)

These chains have a broad, rounded profile. They are relatively slow cutting but stay sharper longer due to the large cutting corner. Leaves a cut of smooth finish. These chains are no longer commonly supplied on modern chainsaws, but can be of use in abrasive cutting conditions. They are somewhat tolerant of less than perfect sharpening techniques.

Full chisel profile (square)

These chains have a square profile. They are fast cutting but, due to the fine working corner, easily blunted in abrasive cutting. Square profile chains are primarily designed to cut green wood. Because of their high cutting speed they are most suited to professional chainsaw users. They require very accurate sharpening, and even small variations from the manufacturer's recommendations can result in reductions in performance and in safety benefits.

Semi-chisel profile (semi-square)

A compromise between round and square profiles. The most versatile of the cutter types, this chain provides the benefit of high performance, reduced vulnerability of the working corner and all of the safety benefits of modern chains. While not as fast cutting as chisel chains, semi-chisel are not as easily blunted. They are widely used, particularly in aboriculture.

Round profile   Square profile   Semi-square profile

Figure 34: Different cutter profiles.

Drive link

The drive link must match the width of the guide bar groove so that the chain exactly fits the bar. The tang, by riding in the bar groove keeps the chain aligned on the bar.

The hook of the tang picks up oil and carries it along the bar for lubrication. The hook also forms a scraper which keeps the bar groove clean.

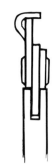

Figure 35: Drive link and bar groove must match.

Gauge

This is the thickness of the tang of the drive link.

The tang is the part of the drive link on which the drive sprocket draws the chain.

Standard gauges are: 0.043 (1.1"), 0.050 (1.3"), 0.058 (1.5"), 0.063 (1.6").

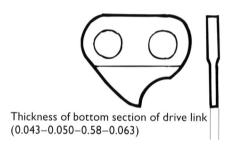

Thickness of bottom section of drive link
(0.043–0.050–0.58–0.063)

Figure 36: Drive link gauge.

Tie straps

The tie strap is the connecting link between the cutter and the drive links, and also acts as a spacing link between the cutters.

Kerf

This is the width of the cut made into the wood by the chain.

The width of the cut decreases as the length of the cutter decreases with wear and sharpening.

In some cases well worn chains will not successfully cut thick, heavy bark due to the reduced width of the kerf.

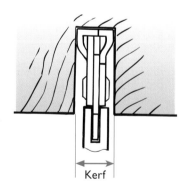

Kerf

Figure 37: Kerf width.

Saw chain sharpening

It is essential that the saw chain is kept sharp. Signs that a saw needs sharpening include:

- you have to apply pressure to make the saw cut
- the cut material is coming out as fine dust rather than 'chip'
- the saw 'dishes' or runs off instead of cutting straight
- damage is present on the chrome surface of top plate or side plate.

Preparation for sharpening

- Don't attempt to sharpen an oily or wet chain. Cut some dry wood to clean the chain.
- File a little bit often, rather than a lot at once.
  The art of sharpening is to 'hone' rather than to remove a lot of metal.
- Make sure the saw is firmly positioned and the bar guide is secured. A stump vice can be used if working in the bush.
- Position the guide bar in the vice so that your wrists are not over the chain. Generally it is best to secure the guide bar near the tip, but without force on the sprocket nose.
- Match file diameter to file guide and pitch of chain (refer to Table 4 below).

Figure 38: Using a stump vice.

Round file diameter

Always use the correct diameter file for the particular chain. File diameter is governed by the chain's pitch. Generally the larger diameter file is used until the cutter is half worn, then the smaller diameter file is used.

Table 4: File diameter and chain pitch.

Chain pitch	File diameter (round file)	Gauge
1/4"	5/32" (4.0 mm)	0.65 mm
0.325"	5/32" (4.0 mm) to 3/16" (4.8 mm)	0.65 mm
3/8" Low Profile	3/16" (4.8 mm)	0.65 mm
3/8" Standard	13/64" (5.2 mm) to 7/32" (5.5 mm)	0.65 mm
0.404" Standard	7/32" (5.5 mm)	0.80 mm

Points to remember when filing cutters:

- Use a file guide recommended by the chain manufacturer.

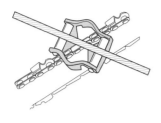

1 = Sighting edges for side plate angles
2 = Sighting edges for top plate filing angles
3 = Cut out for depth gauge setting
4 = Bar groove cleaner and scale for bar groove depth (mm)

Figure 39: Roller file guide.

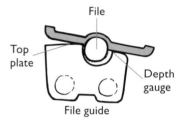

Figure 40: Multifunction gauge.

- Ensure the file guide rests both on the top plate and depth gauge.
- First sharpen the cutter with the shortest top length, then sharpen all other cutters to match. File until all damage is removed from top plate and side plate, i.e. back to unworn plating or chrome surface.
- File from inside to outside using a full length stroke. Since the file will only work when used in the forward direction, hold the file clear of the cutter and do not apply pressure on the return stroke.
- File all the cutters on one side of the chain first, then file the cutters facing the opposite direction.
- File each cutter to maintain the profile of the side plate.
- Try to keep cutters equal in length and angles. Failure to do so will cause chain to cut to one side and excessive vibration.
- Finish each cutter with a crisp leading edge, with no gap between the file and leading edge of the cutter.
- After sharpening the cutters, check each depth gauge for correct clearance.

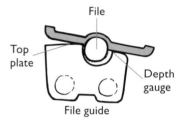

Figure 41: Positioning the guide.

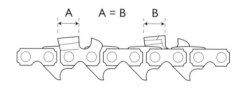

Figure 42: Keep cutters equal.

Top plate angles

Equal top plate angles of both left and right hand cutters is critical for effective cutting.

Side plate angles

Side plate angles also need to be maintained for effective cutting.

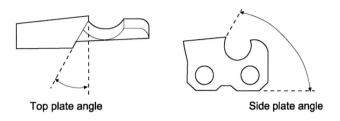

Top plate angle   Side plate angle

Figure 43: Filing angles.

Side plate hook

This is to be avoided as it can lead to an aggressive cutting chain and damage to the drive clutch. It usually results from using a file too small in diameter.

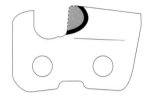

Side plate hook

Side plate back slope

This is to be avoided as the cutter will not cut. It is usually the result of using a file too big in diameter.

Depth gauge setting

Depth gauge setting is critical to the performance of the chain. With wear and sharpening, the top plate becomes lower, therefore it is necessary to also lower the depth gauges to maintain the correct clearance between the cutter and depth gauge. Usually depth gauge profile will require attention about every third to fourth sharpening depending how much of the cutter had to be filed away to repair the cutting edge.

Side plate back slope

The depth gauges on modern chains are designed to reduce the potential for kickback and vibration, therefore it is essential that when lowering depth gauges the original shape of the depth gauge be maintained as near as practical. Always use the depth gauge tool and settings recommended by your chainsaw's manufacturer.

Depth gauge setting varies depending on chain saw pitch and the wood to be cut. Check your manufacturer's recommendations.

Table 5: Recommended depth gauge settings.

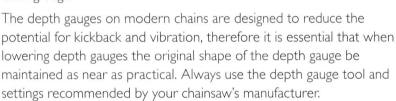

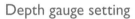 0.65 mm (0.025 in.)	0.025" or 0.65 mm	Softwood Small saws
	0.0250" to 0.030" or 0.80 mm	General cutting Hardwood Large saws

Problems caused by incorrect depth gauge setting are listed in Table 6.

Table 6: Problems caused by incorrect depth gauge setting.

	Depth gauge too low	Increased chance of kickback. Chain grabs, causing excessive bar wear and stretching of the chain. Increases pressure on the drive sprocket and clutch.
	Depth gauge too high	Chain loses self feeding characteristics and extra pressure is required to get the chain to cut. Leads to rapid wear on all components.
	Uneven depth gauge setting	Gives excessive vibration and chain chatter.

Points to remember when filing depth gauge:

- If the depth gauge protrudes above the setting gauge, file it level with a flat file and depth gauge tool.
- As for cutters, work on one side first then the other, always filing from the inside out.
- Make sure all depth gauges are filed to the same height.
- Finally, slightly file each leading edge to round the corner back to its original shape. This will reduce the tendency for kickback.

Depth setting gauge

Rounding leading edge of depth gauge

Figure 44: Depth gauge filing.

Chain tension

Chain tensioning should be carried out when required. Correct chain tension is where the bottom of the chain's tie-straps is just touching the underside of the guide bar. The chain must be able to be pulled around the bar freely using both hands, and should snap back smartly when pulled by the tie straps.

Sprocket nose bars require the chain to be tensioned slightly tighter.

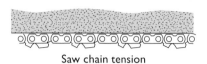

Saw chain tension

Figure 45: Loose chain (top) and correct tension (below).

> ## Remember: a loose chain increases the potential of kickback, and may also snap or derail.

Tensioning a saw chain

1. Loosen bar nuts.
   Make sure you wear cut-resistant gloves.

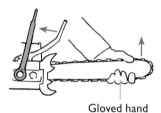

Gloved hand

2. Tighten tension screw.
   Tighten until chain just touches bottom bar rails.
   For sprocket nose bars turn the adjuster one quarter turn more to achieve correct tension.

3. Pull chain around bar in direction of travel to be sure it fits drive sprocket and bar.

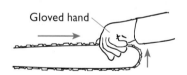

Gloved hand

4. Hold bar tip up. Tighten nuts.

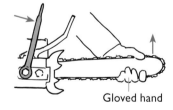

Gloved hand

5. Check that chain moves freely and snaps back when pulled.

Figure 46: Steps in tensioning a saw chain.

Table 7: Cutting attachment troubleshooting chart.

Problem	Possible causes
Chain chatters	• High depth gauges • Badly worn sprocket teeth • Irregularly shaped cutters • Badly stretched chain (more than 6 mm per 30 cm of chain) • Incorrect matching of sprocket pitch and chain pitch
Chain dulls quickly or cannot be successfully sharpened	• Thin feathered top plate (file held too low or pressed down too hard) • Cutters rubbing against the wood (high depth gauges) • Abraded cutters (from hitting rocks, grit, dirt) • In this case the cutters will need to be filed back behind the abrasive mark, and all non-abraded cutters filed back equally. The depth gauges should be lowered accordingly.
Worn or cracked tie straps	• Cracked tie straps at rear rivet holes – caused by back-sloping cutters or blunt cutters • Crack at front rivet hole – caused by high depth gauges, pressure applied to make chain cut • Chain worn straight across the bottom with both rivet holes cracked – caused by tight chain or poor lubrication
Chain breakage	• Heat and excessive friction • Poor sharpening • Improper joining • Excessive wear
Chain pulls or dishes to one side	• Guide bar rails worn unevenly • Incorrectly set depth gauges • Filing angles – either top, or side plates, or right hand cutters – are different from those of the left hand side. Longer cutters on one side. • Abraded cutters, either left or right hand • Spread guide bar rails
Tight joints	• Loose chain • Round bar rails and/or worn satellite tip • Worn cutters, excessive pressure, chain chatter • Worn sprocket • Badly peened rivets of joining links

5. Chainsaw techniques

Pre-start checks

Before starting up your chainsaw, there are some essential checks you must do.

- Are you wearing the correct Personal Protective Equipment (PPE)?
- Warning signs displayed?
- Is the saw chain sharpened and correctly tensioned?
- Sufficient fuel and oil?
- Safety devices attached and operational? Chain brake working?
- All obstructions in the path of the chainsaw removed?
- Working environment as safe as practicable?
- Any other person or machinery in work zone?

Mixing and adding fuel

> **A spark or hot surface can easily ignite petrol or petrol vapours. Make sure you are in a clear area and follow safe procedures.**

Safety precautions

- Always stop the motor and allow to cool before refuelling.
- Place the saw on clear ground or on a firm surface such as a stump away from any dry fuel material. If still in the forest, check for any overhead hazards.
- Don't smoke when handling fuel.
- Take care not to spill fuel on hot components, and wipe any spilt fuel from saw.
- Avoid fuel and oil contacting your skin. It is flammable and a carcinogen.

Figure 47: Refuelling.

Procedure

- Make sure you are using the correct fuel/oil 2-stroke mixture. Check your owner's manual and refer to the chart below.
- Fill fuel and bar oil at the same time. Fill the oil tank first to allow the saw to cool.
- Use chain/bar oil as recommended by manufacturer. (Bar oil is not suitable to mix with fuel).
- Store bar oil and fuel in suitable, clearly marked containers. Do not use glass.
- Take care not to contaminate fuel and oil with sawdust, dirt or water.

Figure 48: Do not start saw near fuel supply.

Calculating chainsaw fuel/oil ratio

When mixing petrol and 2-stroke oil, always refer to manufacturer's recommendations.

When calculating amounts, note that 1 cc = 1 ml, and 1 Lt = 1000 cc.

To calculate the amount of oil to be added to the fuel:

1. Identify the container size (volume in cc)
2. Divide this number by the fuel ratio
3. The answer is the amount of oil required (ml)

For example:

1. Container size is 4 Lt or 4000 cc
2. Fuel/oil ratio required is 25:1
3. 4000 ÷ 25 = 160

Therefore to get 25:1 fuel/oil mix in a 4 Lt container, you need to add 160 ml of 2-stroke oil to the container, then fill with petrol.

The table below sets out the calculations for various container sizes and fuel/oil ratios.

Table 8: Fuel/oil ratio calculator.

Fuel container size	Amount of 2-stroke oil (ml)					
4 Litre	200	160	134	100	80	67
5 Litre	250	200	167	125	100	84
10 Litre	500	400	334	250	200	167
15 Litre	750	600	500	375	300	250
20 Litre	1000	800	667	500	400	334
25 Litre	1250	1000	834	625	500	417
30 Litre	1500	1200	1000	750	600	500
40 Litre	2000	1600	1334	1000	800	667
50 Litre	2500	2000	1667	1250	1000	834
205 Litre (44 gallon)	10250	8200	6834	5125	4100	3417
Fuel/oil ratio	**20:1**	**25:1**	**30:1**	**40:1**	**50:1**	**60:1**
1 Litre = 1000 ml						

Safe operating techniques

The risk of accidents and injury is reduced if you follow safe work techniques.

Starting

- Don't start the saw at the place of refuelling – move at least 3 metres away.
- Make sure the area is safe, with no other people within a 2 metre radius.

> **Drop starting is not an acceptable safe method of starting a chainsaw.**

Method 1: Starting saw on ground

This method is recommended for cold starting, and for heavy or hard-to-start saws.

1. Remove scabbard.
2. Depress decompression button.
3. Ensure ignition switch is on.
4. Apply chain brake.
5. Place saw on ground, guide bar pointing to left.
6. Clear away any obstacles, particularly near tip of guide bar.
7. Kneel on left knee and place the right foot through the rear handle. Left hand is on the front handle with the left arm straight and locked. Keep back straight.
8. Pull the starter cord with the right hand.

Figure 49: On ground method.

Method 2: Using 'leg lock' method

This method is recommended for most situations and starting warm saws.

1. Remove scabbard.
2. Depress decompression button.
3. Ensure ignition switch is on.
4. For warm start, have chainsaw on idle.
5. Apply the chain brake.
6. Grip the front handle with left hand, keeping arm straight.
7. Place rear handle behind right leg with saw body supported on upper left leg.
8. Use right hand to pull starter cord.
9. Maintain straight back, look straight ahead.

Figure 50: Leg lock method.

Use of choke for cold start

A cold chainsaw is started with the use of the choke, releasing to the 'half choke' or 'fast idle' position after the engine has fired once. Another quick pull of the starter cord with choke released should start the engine.

Once started, release the throttle lock to return the engine to normal idle speed.

Lubrication check

Check the bar oil level every time you refuel.

To check chain lubrication before and during cutting, position the bar nose over a light background (tree stump, sawdust, etc.), and run the engine at half throttle. Make sure it throws out an increasing trace of oil.

Be careful not to allow the tip of the bar to contact any surface, as kickback may result.

Keep the oil inlet holes and bar groove open and free from dirt.

Figure 51: Checking for chain lubrication.

Grip

When operating a chainsaw always hold it firmly with both hands. The thumb and fingers of the left hand should encircle the front handle, and the right hand about the rear handle.

A chainsaw is best carried in the left hand with the guide bar rearward and chain brake applied, or the engine stopped. When finished with the chainsaw, turn the ignition switch off and fit the guide bar cover (scabbard).

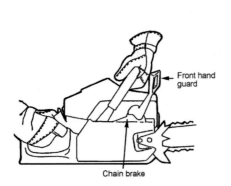

Front hand guard

Chain brake

Figure 52: Gripping and carrying a chainsaw.

Chainsaw reaction

With chain speeds as high as 80–100 km per hour, chainsaws have a natural tendency to react violently when cutting timber. This may cause the operator to lose control and result in severe injury. The operator can minimise chainsaw reaction by maintaining a firm grip and balanced stance, and using the correct engine revolutions when cutting.

Common types of reaction include:

Pull-in reaction can occur when cross-cutting timber with the underside of the chainsaw guide bar. The saw chain draws the saw towards the timber and away from the operator.

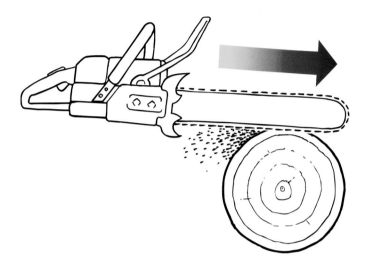

Figure 53: Chainsaw pull-in reaction.

Push-back reaction can occur when under-cutting timber with the topside of the chainsaw guide bar. When making an upwardly directed bottom cut the chain's reactive force will push the saw away from the timber and towards the operator.

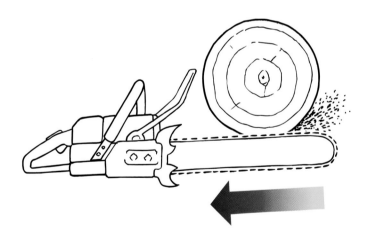

Figure 54: Chainsaw push-back reaction.

Kickback reaction when contact with the tip or upper quadrant of the guide bar nose results in the chainsaw being thrust upwards and backwards in an uncontrolled arc. Kickback reaction generally takes only a fraction of a second to occur.

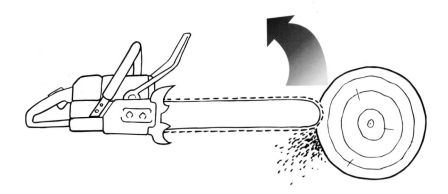

Figure 55: Chainsaw kickback reaction.

When cutting with the chainsaw guide bar in the vertical position, kickback reaction normally results in the chainsaw being violently flung back in the direction of the operator.

In the horizontal position, the reaction will result in the chainsaw 'kicking out' sideways which could be potentially dangerous for bystanders.

Kickback is one of the most common causes of chainsaw accidents. You can help reduce kickback by:

- maintaining a firm grip and balanced stance. The saw must not be in line with your face, but should be slightly to the right of the body.
- cutting at peak revs
- being aware of where the nose of the bar is at all times
- using correct cutting techniques
- making sure the chain is sharp and correctly tensioned
- ensuring correct depth gauge setting
- selecting a narrow profile sprocket tipped bar. This decreases the risk of kickback by reducing the size of danger area in the upper quadrant of the nose. There are also less cutters in the danger area at any one time.

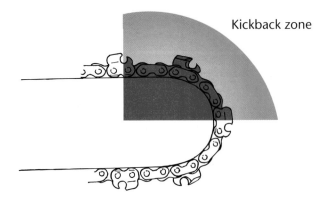

Kickback zone

Figure 56: Kickback reaction zone.

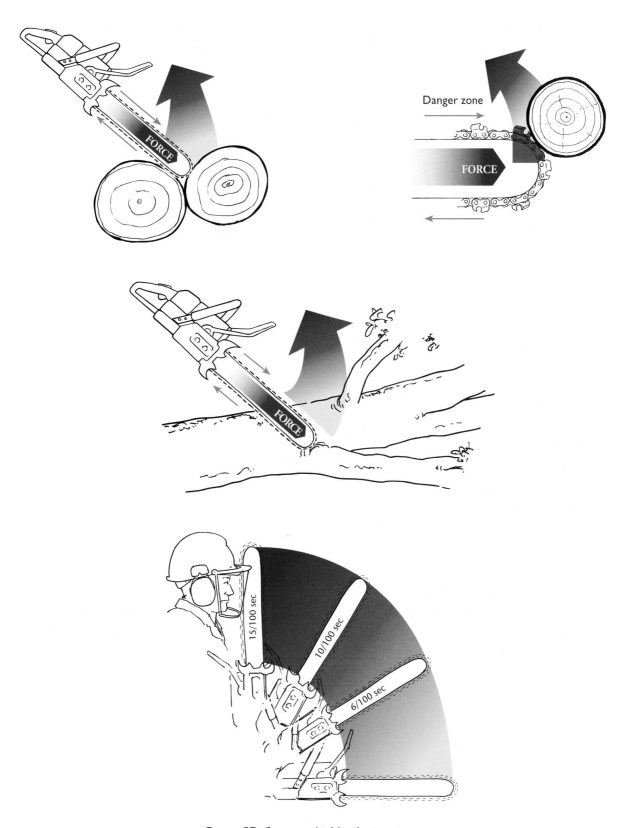

Figure 57: Common kickback situations.

The kickback reaction happens so quickly that you will not be
able to react in time to avoid serious injury, even if the
chain brake works.

Cutting technique

✓ Generally only the upper and lower edges of the guide bar and the lower quadrant of the guide bar nose should be used when cutting. Contact with the tip and the upper quadrant of the guide bar nose should largely be avoided. This will lessen the resultant 'kickback' reaction.

✓ The chainsaw will cut best when not placed under undue force, with the powerhead generally operating at peak revolutions.

✓ Control of the chainsaw must be maintained by having a firm two-handed grip, keeping the powerhead close to the body, keeping a secure footing and a balanced stance on a stable surface.

✓ A chainsaw should only be operated from below shoulder height, never above.

✓ Apply the chain brake before moving branch material, and stop and place the saw on the ground if you need to clear lots of material.

✓ When work is completed, the chain brake should be applied and the chain bar scabbard fitted for transport.

Figure 58: Keep a balanced stance and firm grip.

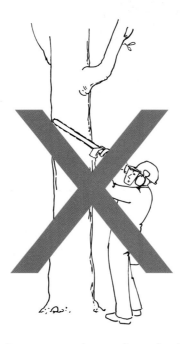

Figure 59: Do not cut with saw above shoulder height.

Vertical cuts

Vertical cuts should wherever possible be made with the rear handle being supported by the upper right leg, and the powerhead stabilised against the forward left leg.

In the vertical cutting position, your eyes should be focused to the left hand side of the cutter bar. Maintain a firm grip and balanced stance, so that should 'kickback' reaction occur, the chainsaw will only be forced backwards and up past the right hand side of the body.

Figure 60: Vertical cutting technique.

Horizontal cuts

Horizontal cuts will require that the left leg in this case be rearward, clear of the guide bar. Control here is aided by the right arm or back of the right hand being supported on the right knee. Adjust your grip on the handles to suit the position of the saw.

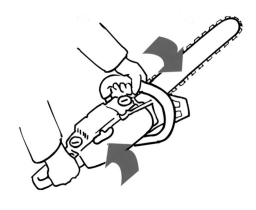

Figure 61: Horizontal cutting technique.

Muscles of the legs should be used in preference to those of the back. So far as possible the back should be kept straight, bending at the knees when cutting close to the ground.

Boring cuts

This type of cut must be done with extreme care because of the risk of kickback. Remember to avoid using the upper quadrant of the guide bar nose.

By operating the chainsaw with a firm grip, locked left arm and well balanced stance at full throttle (peak revs), there should be reduced likelihood of uncontrolled kickback reaction.

Boring cuts are used to:

- assist with cross-cutting materials close to the ground
- cut mortises in fence posts
- open access windows
- test for unsound heartwood in saw logs.

Method 1: Starting with the lower part of the guide bar

This is the preferred method.

Begin the cut using the lower quadrant of the guide bar nose. Once the guide bar is sufficiently working in wood (to a depth of one or two guide bar widths), it should be possible to pivot the chainsaw and then bore it straight into the log.

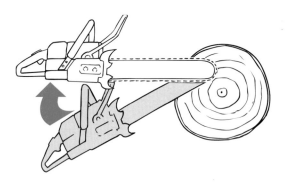

Figure 62: Boring cut preferred method.

Method 2: Starting with the upper part of the guide bar

This method has a higher risk of kickback, and should only be used as a last resort.

Commence cut with the top of guide bar until the depth of the kerf is about the same as the width of the bar. This will serve to guide the bar, effectively eliminating the risk of kickback.

With the saw at full throttle, insert the guide bar in the trunk. Support the saw against your legs.

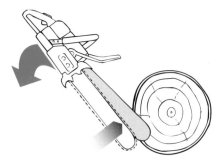

Figure 63: Boring cut (high risk).

6. Trim and cross-cut operations

Limbing

Also known as trimming, limbing is the process by which branch material is removed from fallen trees or logs.

Caution must be exercised when limbing, as there are many potential hazards such as:

- branch material held under tension can whip back towards the operator when cut
- the tree can shift or roll after removal of supporting limbs
- overhead hazards such as dead limbs can drop down.

General basic rules

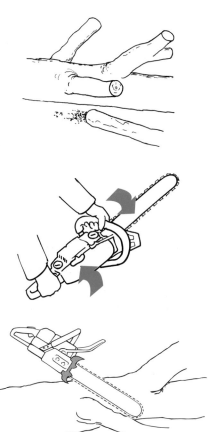

✓	Analyse the forces exerted on limbs. Are they held under tension or compression?
✓	Cut compression wood first, then tension wood last.
✓	Stand in a safe working position and watch out for obstacles.
✓	Where possible never limb on the side of the tree you are standing.
✓	Concentrate on what you are doing.
✓	Always be aware of where the nose of the guide bar is. Beware of kickback!
✓	Do not leave branch stubs.

At times it may be necessary to work the saw horizontally. Adjust your grip on the chainsaw's handles to suit the position of the saw.

Whenever possible, let the tree support the weight of the chainsaw. Pivot the saw, using the saw's dogs (spikes) as a fulcrum. Where this is not possible, support the weight of the saw with your thigh.

Figure 64: Basic limbing techniques.

Assessment of stress in limbs/branches

Timber is composed of fibres which are rectangular in shape, with the long side of the fibre going vertically up the tree. Some trees have long, skinny fibres which are weaker than other trees that have shorter, stronger fibres.

Because of the force of gravity, the fibres are placed under internal stress according to their position in the tree. Fibres may either be:

- under compression (being squashed)
- under tension (being stretched)
- neutral.

For example, a branch extending from the tree is being pulled down by the force of gravity. The branch resists the downward pull of gravity, which places the branch fibres under stress. The underneath side of the branch is under compression, the upper side of the branch is under tension, and the middle of the branch is neutral.

Once you start to cut the branch you change the internal stresses on the fibres. Whether you cut from the bottom or the top, you will reduce the amount of neutral fibres so that more fibres are now under tension or compression. Fibres under compression will store energy depending on the weight of the branch. When the branch is cut, or breaks under the stress, this compressed energy is released and can cause the branch to whip back violently towards you.

To prevent this, you should cut through the compression wood first, slowly, until the branch starts to move. This means there is no neutral fibre remaining. Then cut the tension wood.

If the branch is supported at both ends, the bottom side of the wood is under tension and the top side is under compression.

Hanging branches

Hanging branches (supported at one end only) attached to the trunk should be first undercut, no deeper than about one-third timber diameter, and then severed with a top cut. In large and heavy branches, the cuts may be staggered or 'stepped', the final top cut being placed closest to the supported timber, for better chainsaw control.

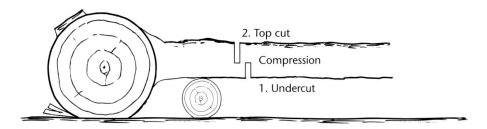

Figure 65: Method of removing a heavy horizontal limb supported at one end only.

Sprung branches

'Sprung' branches (supported at both ends) in contact with the ground must be first top cut in compression, then released by cutting the underside tension wood.

> **Beware!! The tree may be resting on the limbs and roll when the limbs are cut.**

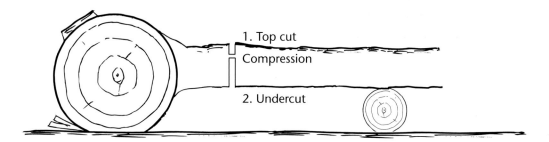

Figure 66: Method of releasing a branch supported at both ends.

When removing the last supporting limbs it may be necessary to stand on the safe side and have the tree chocked for added support, just in case it should roll.

Generally the highest standing limbs should be removed first, without operating the chainsaw above shoulder height.

Wherever possible always try to 'clear as you go'. Failure to provide a clear working area causes the operator to over reach when trimming and compromises chainsaw control.

It is essential that limbs are trimmed completely with flush final cuts, not only because of log quality control requirements, but also for reasons of operator safety.

Typical trimming work patterns

Hardwood limbing

Large spreading fallen trees, especially hardwood species, with much crown held off the ground, present a hazardous situation that requires careful assessment.

The fallen tree must be firmly supported and unable to roll or move while such work is being carried out. This may be aided by removing sections off the butt prior to limbing.

Each limb may need to be sectioned individually.

An example is shown in Figure 67.

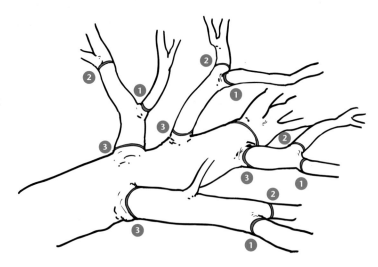

Figure 67: Hardwood limbing, e.g. Eucalyptus, showing cutting sequence.

As each limb is removed, the situation may need to be re-assessed on the merit of the next limb. This operation requires much attention to avoid:

- kickback
- the guide bar becoming pinched
- the saw being thrown back toward the operator as the tension on the limb is released
- operator injury from branches springing back or dropping down, or by the tree shifting. These actions can occur with tremendous force.

The following rules should be observed:

- Carefully plan the cutting of limbs and re-assess according to each individual situation. It is very important to evaluate the tensions in the wood, particularly on large branches. You must also assess the likely movement of the log when removing a limb.
- First cut and remove branches hindering your work.
  Cut branches in two or more sections when there is a danger of cracking at the base or when this facilitates clearing the work area (see Figure 67).
- Keep your working space clear of cut branches.
- Take care to avoid springing limbs when tension is released. If possible stand on the other side of the log when cutting limbs under pressure. If you must stand on the same side because of the log size try to reduce the pressure on the limb by cutting sections off the limb from the small end.

Softwood limbing

When trimming fallen softwood species, such as conifers, it is usual to work from the butt towards the crown. This is due to the regular whorled distribution of branch wood.

Be particularly careful when trimming softwood, as the longer fibre structure means these timbers are often a 'storehouse' of significant physical energy under load. When the timber is cut this energy may release suddenly and violently with considerable force.

A suitable systematic method of dealing with medium-sized trees is the so-called 'six point' or 'lever method' system.

Operating from the left hand side of the fallen tree the chainsaw is efficiently swept back then forth about the trunk in a lever-like manner, removing a whorl of branches in each direction.

The saw moves from the far side to the near side on the first ring (1, 2, 3), and then moves to the next ring, this time cutting from near side to far side (4, 5, 6).

Whilst moving along the log, the weight of the chainsaw is supported by the tree as much as possible.

Both top and underside of the chainsaw guide bar are used when trimming using this method.

When the tree lies on the ground, stand on the left side and work your way to the top.

The tree can then be turned and the remaining branches cut while the operator returns to the butt of the tree.

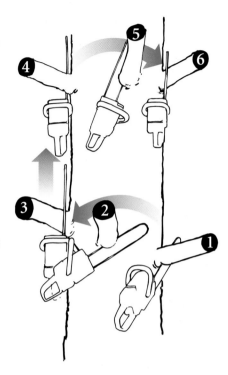

Figure 68: Six point or lever method.

Beware of kickback when using this technique.

When the tree lies across a hollow or there is clearance under the trunk, the branches on the underside of the two rings can be cut in one movement before the operator moves forward to the next two rings.

Figure 69: Softwood limbing.

Basic principles of cross-cutting

Cross-cutting is the process by which trimmed timber is cut to required length, permitting removal in such forms as saw logs and firewood billets.

Follow the basic principles of cross-cutting listed below, to minimise the potential dangers of 'kickback' and log material rolling towards the operator.

- **If not sure of the reaction, do not cut**. Seek advice or assistance.
- Always assess the bind relationships in the log (internal tension in timber) and choose sequence of cuts to suit.
- Stand to one side of cut and on the opposite side to which a log will spring or roll.
- If there is a chance of either half of the log springing, keep an escape route open.
- Wherever possible cut the opposite side of the log first. This will keep the operator as far away from the log as possible when the release cut is made. This will minimise the chance of saw jamming.
- To minimise up-cutting (it's hard work), boring in and down cutting may be easier.
- Keep the saw horizontal to prevent the chain hitting dirt or rocks and the operator from being cut. (A vertical cut through a log could put the operator's legs at risk.)
- Watch kerf to see whether it is opening or closing. Be prepared to alter the sequence of cuts if you find you have misjudged the initial assessment of tension or compression.
- If there is a risk of log pinching or jamming the guide bar, pull the saw back and re-enter the cut (perhaps with a boring cut).
- 'Sawing' the chainsaw back and forth is effective when cutting badly split timber as this widens the kerf of the cut.
- Use the saw as a lever and the logs as a pivot point to minimise work effort. Similarly use your leg muscles when cutting upwards.
- Insert a wedge in the cut if there is a high risk of the cut jamming or the log dropping or twisting.

Figure 70: Always stand on the uphill side of the log.

Cross-cutting techniques

Successful cross-cutting is very much dependant upon the operator recognising how the forces of 'tension' and 'compression' may act within the timber. These are also called 'bind' relationships.

In cross-cutting always cut the wood in compression first, whether that is on the top, bottom or side of where the cut is to be made.

Generally initial cross-cutting cuts in compression wood should be no deeper than about one-third of timber diameter, before making release cuts in tension wood or driving wedges to prevent kerf closure. Failure to do so may cause the kerf to close up, causing the chainsaw cutter bar to become wood bound and jam.

When cross-cutting it is also important to keep the saw level or horizontal to ensure that the saw chain does not strike the ground. Contact with soil or stone will result in a blunted chain, which will need sharpening before cross-cutting can resume.

Trunk sections supported at both ends (bridging cut)

When supported at both ends, trunk sections should be initially top cut in compression, then cross-cut with a final undercut to the tension wood.

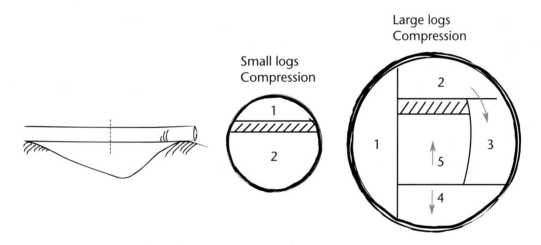

Figure 71: Cross-cut trunk section supported at both ends.

Bridging cut using a wedge

In this technique, cut timber in one downward action, putting in a wedge when clear of the guide bar. Make sure the cut is deep enough to avoid the wedge contacting the saw chain.

Figure 72: Bridging cut using a wedge.

Trunk sections supported by one end only (swinging cut)

Trunk sections supported at one end only must first be undercut in compression wood, then released with a final top cut to the tension wood.

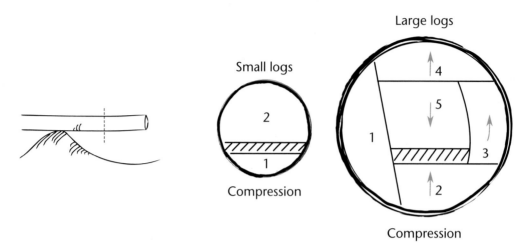

Figure 73: Hanging trunk section supported at one end only.

Trunk sections supported along entire length

The technique below can be used when it is difficult to access the underside of the log, for example on a bitumen road.

Small logs

Where lying on the ground, supported along the entire length, trunk sections that are capable of being rolled may be top cut, then rolled, enabling the cross cuts to be completed. This reduces the possibility of striking the ground with a running chainsaw. A cant hook may be used to assist in turning the timber.

Large logs

Heavy logs lying on the ground can be first top cut, then have a wedge inserted to prevent kerf closure before completing the cut. Make sure you cut deep enough so the wedge will not contact the saw chain. The wedge can be driven in progressively to prevent the kerf closing.

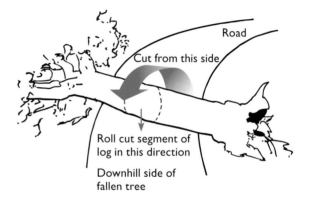

Figure 74: Cross-cutting a small log.

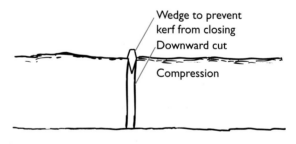

Figure 75: Cross-cut large log lying on ground.

One side in compression (side bind)

Where a log lies across the slope with a side unable to move because of stumps or other logs, side bind can occur. While standing on the compression side of the log, cut the compression wood first but no more than one-third of the diameter of the log with a slightly inward and upward direction, then reach over the log and cut the tension wood second.

Be aware that the log may move quickly away from you as you cut the tension wood and potentially take your chainsaw with it and also pull you off balance.

Cutting logs with side bind can be extremely dangerous and should be done only after all other options have been exhausted, e.g. removal by a machine.

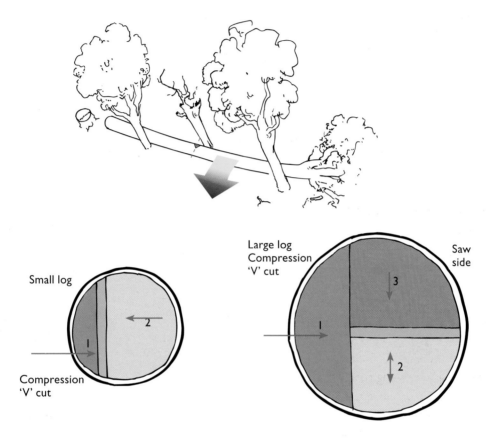

Figure 76: Cross-cutting of log with side bind.

Safety notes:

1. It may be very dangerous to attempt to cross-cut a large diameter tree that is lying under significant side bind tension. Side bind tension should be progressively released by removing the tree's head before attempting the required cut.

2. When putting in the final cut, the operator should stand on the opposite side to the expected movement of the log.

3. The greatest danger is from smaller logs because these are more flexible and therefore more likely to spring back.

Plunge cutting

This is a variation of the bridging cut employed to preserve timber quality for mill/saw logs by preventing the log from splitting. This technique may be used in conjunction with reducing cuts, lessening the overall diameter of large logs for cross-cutting.

Following identification of the compression point at which kerf closure is about to occur, a number of plunge cuts are made back and forth into the log prior to final downward cutting of the wood held in tension.

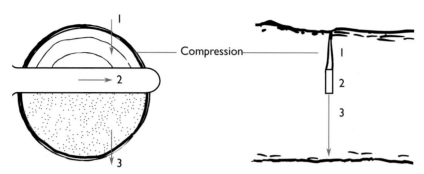

Guide bar length greater than log diameter

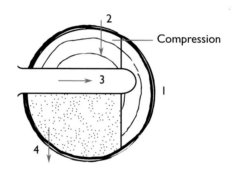

Guide bar length less than log diameter

Figure 77: Plunge cutting technique – method of cross-cutting for log quality.

Log trap

Before cutting, check the log for any possible lateral movement. If one end of the log is unable to move and the other end can drop away when cut, an angled cut may be used. The angle must be sloped to the correct side so that the release cut will not jam the saw when one end drops.

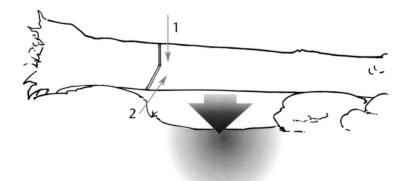

Figure 78: Angled cut through a log trap.

Wind throw and pushed trees

When cutting trees that have been blown over by wind or have been pushed by a machine, operators should consider the following:

- Uprooted trees are potentially dangerous as the trunk is subject to extreme internal stress, and the root plate might stand back up into the stump hole when the trunk is cut. Machinery may be necessary to deal with wind-thrown trees.

Figure 79: Beware of stump springing back.

- Always stand clear of the exposed root mass. Try to cross-cut as closely as possible to the exposed root system, to lessen any stump movement. The root mass can be propped up and the hole backfilled to prevent the roots from falling back when the trunk is cut.
- First make an undercut into the compression wood, followed by a top cut in the tension wood. Proceed cautiously to avoid a jammed guide bar or 'thrown' chainsaw.
- Stagger the bottom and top cuts, with the top cut to the crown end of the trunk.
- Use minimal length of guide bar when making the cuts, thereby keeping your body clear of the wind-thrown trunk material.
- If the tree is laying downhill or has been pushed out of its stump hole, then it is likely that, when cut, the stump will fall forward towards the barrel of the tree. In this instance the cuts would be placed in the tree as for a bridging cut, making sure to cut the log far enough from the stump to stop the stump falling forward enough to cause injury.
- To reduce the likelihood of sudden movement by the stump, operators can trim limbs and cut sections from the head of the tree using the method outlined in the hardwood trimming section of this manual.

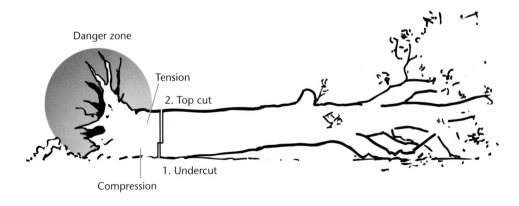

Figure 80: Place stagger cut as close as possible to root mass.

Butt trimming

> **Butt trimming is a very dangerous activity due to the high risk of kickback.**

Butt trimming or pre-limbing of trees may be necessary to remove branches that may interfere with the felling operation. This is particularly so with softwood felling.

Basic principles

- Don't cut branches above shoulder height. Saw will have a strong tendency to kickback and physical control is poor.
- Use only a downward cutting action, and work around tree in an anti-clockwise direction.
- Hold the saw so that the guide bar is at right angles to your body – in this way, if the saw does kickback there is little risk of it striking you. (Keep out of the kickback line during any chainsaw operation where kickback risk is high.)
- Face visor is essential. Risk of eye injury is high due to flying chips.

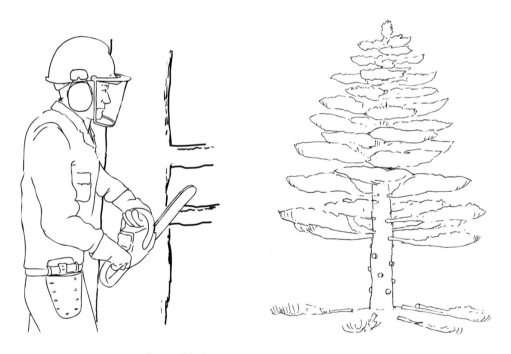

Figure 81: Butt trimming a softwood tree.

Ripping

Ripping is the lengthwise cutting of timber in the same direction as the wood grain.

Only highly powered professional chainsaws are capable of effectively ripping hardwood fencing materials.

Do not straddle the timber when ripping. It is a much safer practice to stand to one side.

When ripping, ensure that the timber is properly secured so that it cannot roll. Use chocks and branch material to keep half ripped logs off the ground, so that the saw chain does not come into contact with the soil and become blunted whilst making final ripping cuts.

Operating the chainsaw at peak revolutions, draw it along the log in a lever-like manner with the aid of the chainsaw bumper spikes. These should rest on the log so that the chainsaw is supported by the timber rather than carried by the operator.

Figure 82: Stand to one side when ripping.

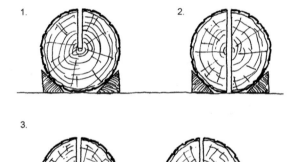

Figure 83: Hardwood timber ripped into four fence posts.

Scrub clearing

At times it may be necessary to use a chainsaw in the horizontal cutting position for clearing scrub. For example, this might be required to establish a clear work area when operating in heavy undergrowth.

When scrub clearing, maintain a well-balanced stance. Bend your knees and keep your back straight. Have the chainsaw powerhead in contact with your body, perhaps the inside of your right thigh with the right leg in the forward position.

Work systematically, using the chainsaw in a sweeping motion from right to left. Keep the saw as low to the ground as is practical. Leave the cut stems with flat tops, not points, to lessen the possibility of injury in case of slips or falls. Stems with flat cut tops are also less likely to damage tyres of vehicles used in fire fighting or other forestry operations.

Always be mindful of avoiding contact of the 'kickback zone' of the guide bar nose with stems. This can cause a potentially dangerous 'kickout reaction' of the chainsaw.

7. Relevant state authorities and technical standards

Regulatory and licensing requirements for chainsaw operators vary from state to state. Check with your local state authority listed below:

Tasmania

Regulatory Authority:

Workplace Standards Tasmania
Level 3 Henty House
1 Civic Square
Launceston Tas 7250
Phone: 03 6233 7657
www.wst.tas.gov.au

Licensing Authority:

Tasmanian Forest Industries Training Board
Shop 4
Cornwall Square Transit Centre
Cimitiere Street
Launceston Tas 7250
Phone: 03 6331 6077

Northern Territory

NT WorkSafe
www.worksafe.nt.gov.au

First Floor
Darwin Plaza Building
41 Smith Street
The Mall
Darwin NT 0801
Phone: 1800 019 115

Ground Floor
Government Centre
First Street
Katherine NT 0851
Phone: 08 8973 8416

Peter Sitzler Building
67 North Stuart Highway
Alice Springs NT 0871
Phone: 08 8951 8682

Victoria

WorkSafe Victoria
222 Exhibition Street
Melbourne Vic 3000
Phone: 1800 136 089
www.workcover.vic.gov.au

New South Wales

WorkCover NSW
92–100 Donnison Street
Gosford NSW 2250
Phone: 02 4321 5000
www.workcover.nsw.gov.au

Queensland

Workplace Health and Safety Queensland
75 William Street
Brisbane Qld 4000
Phone: 1300 369 915
www.deir.qld.gov.au/workplace/

South Australia

Workcover SA
100 Waymouth Street
Adelaide SA 5000
Phone: 13 18 55
www.workcover.sa.gov.au

Western Australia

WorkSafe WA
5th Floor, Westcentre
1260 Hay Street
West Perth WA 6005
Phone: 1300 307 877
www.commerce.wa.gov.au/WorkSafe

Technical standards

AS 2727:1997	*Chainsaws – Guide to Safe Working Practices*
AS 2726.1:2004	*Chainsaws – Safety Requirements, Part 1: Chainsaws for General Use*
AS/NZS 4453.3:1997	*Protective Clothing for Users of Hand-held Chainsaws, Part 3: Protective Leg Wear*
AS/NZS 1801:1997	*Occupational Protective Helmets*
AS/NZS 2210.1:1994	*Occupational Protective Footwear, Part 1: Guide to Selection, Care and Use*
AS/NZS 1337:1992	*Eye Protectors for Industrial Applications*
AS/NZS 1336:1997	*Recommended Practices for Occupational Eye Protection*
AS/NZS 1319:1994	*Safety Signs for the Occupational Environment*
AS/NZS 1270:2002	*Acoustics – Hearing Protectors*
AS/NZS 4602:1999	*High Visibility Safety Garments*